I0816509

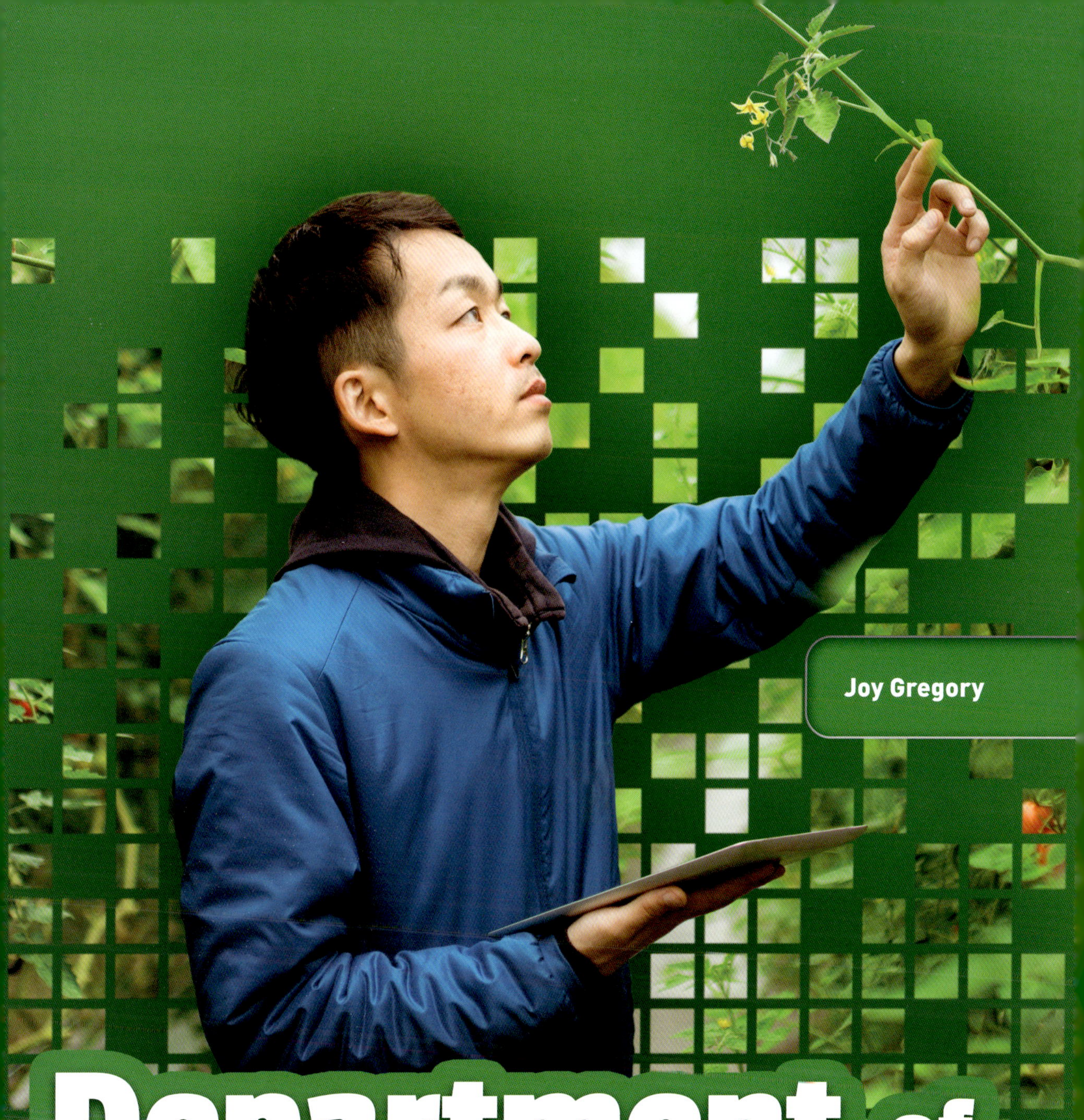

Joy Gregory

Department of Agriculture

POWER • AUTHORITY • GOVERNANCE

LIGHTBOX
openlightbox.com

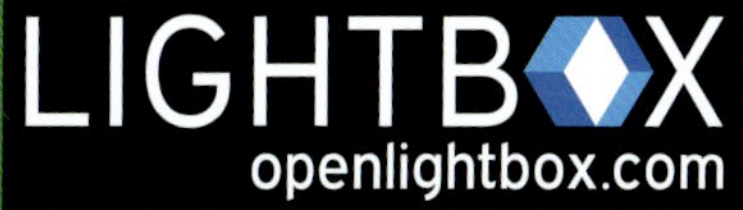

Go to
www.openlightbox.com
and enter this book's
unique code.

ACCESS CODE

LBXB4979

Lightbox is an all-inclusive digital solution for the teaching and learning of curriculum topics in an original, groundbreaking way. Lightbox is based on National Curriculum Standards.

LIGHTBOX SUPPLEMENTARY RESOURCES

SHARE
Share titles within your Learning Management System (LMS) or Library Circulation System

CURRICULUM
Find national and state curriculum correlations

CITATION
Create bibliographical references following APA, CMOS, and MLA styles

STANDARD FEATURES OF LIGHTBOX

AUDIO High-quality narration using text-to-speech system

ACTIVITIES Printable PDFs that can be emailed and graded

SLIDESHOWS Pictorial overviews of key concepts

VIDEOS Embedded high-definition video clips

WEBLINKS Curated links to external, child-safe resources

TRANSPARENCIES Step-by-step layering of maps, diagrams, charts, and timelines

INTERACTIVE MAPS Interactive maps and aerial satellite imagery

QUIZZES Ten multiple-choice questions that are automatically graded and emailed for teacher assessment

KEY WORDS Matching key concepts to their definitions

This title is part of our Lightbox digital subscription

Lightbox Grades 3–5 Subscription
ISBN 978-1-5105-5424-5

Access hundreds of Lightbox titles with our digital subscription. Sign up for a **FREE** subscription trial at **www.openlightbox.com/trial**

POWER • AUTHORITY • GOVERNANCE

Department of Agriculture

CONTENTS

Introduction

The United States Department of Agriculture (USDA) is the part of the government that is responsible for food and agriculture. The department is divided into separate agencies that work together. Each agency deals with different, specific issues.

Several USDA agencies oversee how food is grown. Some of these focus on farming and ranching. Others promote the **export** of food or help farmers get **loans**. The USDA's agencies make sure that food is safe to eat and is properly labeled. The department also runs the Forest Service. This agency protects the country's forests and grasslands.

The USDA does a large amount of scientific research. This research helps farmers and ranchers grow more food. Since people need healthy food to eat, some of this research is about food and **nutrition**. The USDA also does market research. The information that this research provides helps farmers sell their crops.

The USDA's headquarters are located at 1400 Independence Avenue, in Washington, DC.

Most of America's food and fiber, which comes from crops and animals, is grown in **rural** parts of the country. USDA agencies help support people who live in these areas. One helps rural residents buy houses. Another program builds important utilities, such as water and waste treatment plants.

The United States is governed by the **Constitution**. Its citizens give their power and authority to the government. In return, the government serves the will of the people. This is called "popular sovereignty." As an example of popular sovereignty, every USDA law must obey the Constitution.

USDA employees work in **4,500 locations** around the world.

The USDA **employs** about **100,000 people** in dozens of **offices** and agencies.

The United States **exports** about **$139 billion** worth of **food** and **agricultural products** each year.

Origins of the USDA

Agriculture has always been important to U.S. citizens. In the 1700s, Americans grew much of the food they ate. Then, they sold anything extra to their neighbors. In the 1830s, Congress made agriculture part of the U.S. Patent Office. This gave agricultural issues a place in government.

President Abraham Lincoln created the USDA in 1862. At the time, about half of the country's population was **agrarian**. Lincoln himself was a supporter of agriculture. He grew up on a farm in Kentucky. Lincoln praised inventions that made farming easier, and helped improve education for farmers and ranchers.

Over time, the USDA grew. By 1914, the department was helping to fund farming classes. New agencies were added in response to the **Great Depression** in the 1930s. These aided struggling citizens in several ways. One provided loans, helping farmers buy essential products such as livestock feed. A separate USDA agency was created to help feed the hungry. It operated throughout the United States.

In the 1950s and 1960s, USDA research increased. Scientists helped farmers grow new crops and taught farmers how to grow more food while using less land. The need for better food information also became clearer to the U.S. government. In 1981, the Food Safety and Quality Service of 1977 was renamed the Food Safety and Inspection Service. It was responsible for tracking down food that could make people sick.

The USDA was founded during the American Civil War. President Lincoln considered agricultural development a key policy that could not wait until the war's end.

Branches of Government

The U.S. federal government has three branches. Each branch acts as a check on the power of the others. This balances the power each branch holds, creating a system of "checks and balances."

The USDA is in the **executive branch**, which carries out laws. This branch includes the president, vice president, and department leaders. Each department leader is appointed by the president. Together with the vice president, they form the cabinet. The cabinet advises the president.

The federal government also has a **legislative branch** and a **judicial branch**. The legislative branch is Congress. It controls the **budget**. Congress can cut funds to check the USDA's power. The judicial branch is the country's legal system. It resolves conflicts relating to U.S. laws, preventing departments from going beyond their authority.

Purpose of the USDA

The USDA supports the food industry in the United States. Food and agricultural businesses contribute $1.1 trillion to the U.S. economy each year. They also directly employ about 22 million U.S. citizens. This includes people working in food factories, restaurants, grocery stores, and those that drive food transport trucks. The country's biggest food exports are soybeans and beef.

About 2.6 million Americans work on farms and ranches.

Most of the USDA's budget is spent on food aid. This money is used to feed **food-insecure** people in the United States. The Supplemental Nutrition Assistance Program (SNAP) helps people buy food. Other programs provide meals for children in schools.

The USDA also works to protect future food supplies. To do so, it runs programs that help the environment. About half of all the land in the United States is used to grow food. USDA scientists work on ways to reduce this amount. They also help farmers find ways to grow food with less water. USDA scientists work to keep agricultural waste out of the water, making it safer for the environment.

California, Iowa, Nebraska, and Texas grow the most food of any U.S. states.

The USDA runs more than 50 of its programs in rural areas. These programs improve life in these locations in many different ways beyond simply helping people produce crops. One such way is providing health care equipment, such as ambulances. Another way is by loaning money to invest in the expansion of internet services throughout the country.

SNAP

Hunger was a national problem during the 1930s. The U.S. government tried to address the issue in several ways. In 1969, the USDA set up an agency to deal with hunger, called the Food and Nutrition Service (FNS). It runs 15 programs.

The biggest FNS program is SNAP, formerly known as the Food Stamp Program. Many of the people SNAP helps have jobs, or live with parents that have jobs, but do not have enough money to buy healthy food. In 2021, SNAP provided assistance to about one in every nine U.S. citizens. In total, this was about 41.5 million people. Of these people, almost 44 percent were children, while millions of others were disabled adults.

USDA Through the Years

The USDA was created more than 150 years ago. Much has changed since that time. Today, less than 2 percent of the U.S. population lives on farms and ranches. In spite of this, the United States remains a world leader in food production, making the USDA more important than ever before.

May 15, 1862

President Abraham Lincoln signs the law that creates the USDA.

1887

The Hatch Act of 1887 passes. It creates farm research stations in every state.

February 9, 1889

The USDA receives cabinet status and joins the executive branch of the federal government.

1923

The USDA opens the Bureau of Home Economics. The bureau researches and tests practices relating to life skills, such as cooking, storing food, and safely storing household goods.

1935

Widespread droughts hit the Great Plains region. The USDA establishes programs intended to provide ranchers with loans or help move them to drought-free places.

1939

World War II begins. In response, the USDA collects data about food production and supplies across the country.

1941

The Chicago Dairy Office opens. It helps the USDA tracks dairy production.

August 8, 1969

The USDA opens the FNS. It provides food to hungry people.

July 23, 1972

The Landsat 1 satellite is launched. Today, the Landsat program is used to collect information about U.S. crops.

1985

The Conservation Reserve Program begins. It protects land that includes important wildlife habitats.

1991

The USDA surveys chemical use on farms. This data is shared with consumers around the nation to provide them with information about how their food is grown.

1995

The USDA creates its first website. The site posts USDA reports minutes after they are created.

2008

The USDA reports its first statistics about **organic** agriculture.

2020

The global COVID-19 health pandemic hits the United States. The USDA buys U.S.-grown food for food banks to help those unable to work because of the pandemic.

USDA Issues

The USDA protects the U.S. food supply. The department's programs help farmers grow food and make sure people have food to eat. However, the USDA has also faced several controversies throughout its history as it has tried to accomplish these goals.

One major issue facing the USDA is the growth of the **genetically modified** food industry. Genetically modified crops include soybeans, corn, and sugar beets. Many of the animals that people eat feed on these crops as well. The USDA has approved many of these foods, and meat from animals that eat these foods, for humans to eat. Some people support genetically modified crops. The plants may grow larger, more quickly, and be more disease-resistant than those that are not modified. However, others are concerned that these foods may not be safe to eat over a long period of time. They are also worried that genetically modified crops could contaminate organic foods.

Other concerns relating to the USDA include the types of food available through some of its services. People debate whether the USDA should only provide healthy options, or if people should have more choices. In 2010, New York City tried to stop people from purchasing soft drinks under SNAP. However, the USDA denied the city's request.

The USDA works with two other U.S. government agencies, the U.S. Food and Drug Administration (FDA) and the U.S. Environmental Protection Agency (EPA), to ensure that genetically modified crops are safe. The USDA sets rules to ensure that modified crops cannot harm other plants through the USDA Animal and Plant Health Inspection Service (APHIS).

In early 2020, the USDA proposed changes to the National School Lunch Program's (NSLP's) rules. The department said that these changes would make the program more flexible and give schools more choice about the foods they served. However, many people felt that these changes would mean that children in schools would end up eating fewer healthy foods, such as green vegetables.

Pigford v. Glickman

The USDA provides loans and assistance to many people. This money can be used to buy, operate, or improve a farm. Under the U.S. Constitution, every borrower should be treated the same.

In 1999, a U.S. court settled a **civil rights** lawsuit known as the Pigford Case. This lawsuit argued that the USDA had treated African American farmers unfairly. Between 1981 and 1996, some of these farmers were denied loans, or had their loans delayed. Many of these farmers went bankrupt. They could not pay their bills, so the banks took away their farms. The court awarded money to more than 30,000 African American farmers. Later, in 2010, Native American farmers won a similar case, Keepseagle v. Vilsack.

When the Pigford Case was concluded, it was the largest civil rights settlement in U.S. history at the time.

Key Figures in the USDA

Today, the person in charge of the USDA is called the secretary. Before 1889, that person was called the commissioner. The USDA has been led by 6 commissioners and 31 secretaries over its history.

Isaac Newton

Isaac Newton (1800–1867) was born in New Jersey. He worked for presidents Abraham Lincoln and Andrew Johnson. In 1861, Newton became the first superintendent of the U.S. government's agriculture division. He became the first USDA commissioner the following year.

Ann M. Veneman

Ann Veneman (1949–) was the 27th secretary of the USDA. She served from 2001 to 2005. A lawyer, Veneman is the first, and so far only, woman to hold this title. Veneman grew up on a peach farm in California. Her career was highly focused on food issues.

Tom Vilsack

Tom Vilsack (1950–) served as USDA secretary under president Barack Obama. In his first term, Secretary Vilsack targeted childhood hunger. He also helped improve the safety of the U.S. food supply. Vilsack was sworn in again as secretary of the USDA in early 2021.

Childhood Hunger

HISTORICAL CASE STUDY

Childhood hunger is a problem in the United States. Across the country, one in every seven children is considered to be food insecure. Most of the parents of these children are considered food insecure as well.

Healthy food benefits children. It helps their brains develop and their bodies grow. Healthy food also prevents diseases and helps children learn. The NSLP was established by President Harry Truman in 1946. Today, the program helps almost 20 million children to get free meals at school.

In 2010, Congress passed the Healthy Hunger-Free Kids Act. This act gave money to schools that served healthy meals. It set rules about the food sold in schools. The act also made it easier for children to get school meals. It was intended to provide about 21 million more meals a year to schoolchildren.

President Truman established the NSLP in part because, during the 1940s, many young men who attempted to join the U.S. military were unable to due to conditions caused by poor diets growing up.

Careers in the USDA

Food Inspector

USDA food inspectors study USDA rules. They are trained to spot food safety problems. Food inspectors often inspect places where fruits and vegetables are picked and packaged, along with meat and poultry plants. They also visit grocery stores and restaurants where food is prepared. Inspectors can fine people or businesses if they break food safety laws.

Crop Scientist

USDA crop scientists study food plants grown in the United States. They work in fields and laboratories. Crop scientists write and publish reports that can help solve many problems. Some crop scientists breed plants that grow faster. Others develop plants that taste better. Different crop scientists study insects or diseases that harm plants.

Survey Statistician

Survey statisticians work with numbers. Many work for the National Agricultural Statistics Service (NASS). NASS surveyors ask farmers questions and gather data about crops, livestock, and the weather in rural areas. Statisticians use this data to help farmers and ranchers decide what they should grow.

Dietitian

Dietitians are food experts who study nutrition, or how food affects health. USDA dietitians write about nutrition. They design educational programs for people of all ages. Some programs teach children about healthy food. Others teach adults how to shop for or prepare healthy foods.

Demand for agriculture and **food scientists** is increasing. Between **2020** and **2030**, the number of people in this field is expected to **grow** by **9 percent**.

Close to **70 percent** of NASS employees are **math experts**. Another **11 percent** are **Information Technology (IT) specialists**. They oversee computer systems.

AGRICULTURE COUNTS

Most **USDA careers** are in fields linked to **Science, Technology, Engineering,** and **Math** (STEM). Those careers include jobs at more than **90 USDA Agricultural Research Service sites**.

Tools of the Trade

Workers with the USDA use many different tools on the job. A large number of these tools are used to collect information. That data supports food production throughout the nation.

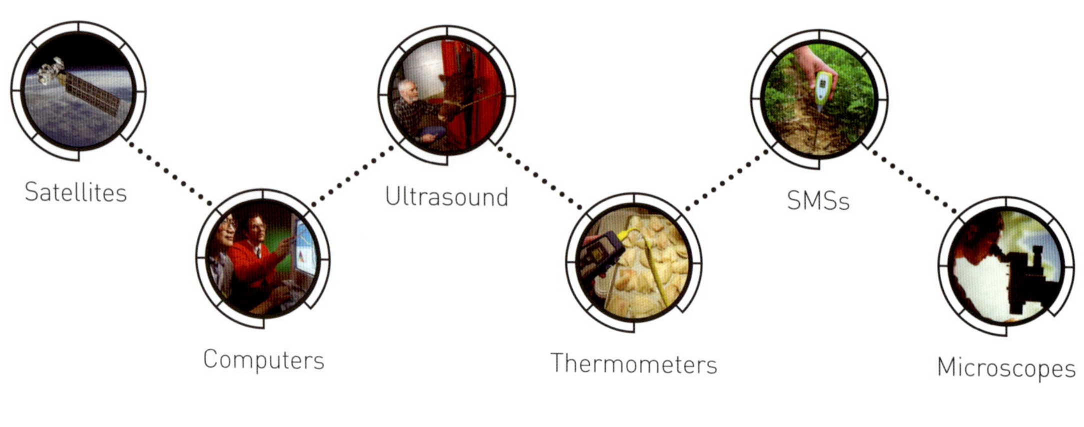

Satellites

Satellites are objects that orbit Earth. Some have cameras that can collect pictures of crops from hundreds of miles (kilometers) overhead. The USDA uses these pictures to identify crops with insect problems, flood damage, or disease. This information helps people to predict crop yields. It also shows areas where farms may need government assistance.

Computers

The USDA collects a large amount of information. Workers use computers to gather and store this data. Computers are also used to compare information about crops and livestock. They help the USDA track disease outbreaks and share foreign sales data. The USDA also uses computers to analyze information. This helps the department identify problems quickly.

Ultrasound

Ultrasound machines are used to "look" inside the body of a living thing. They use sound waves to form pictures of organs and tissues. The USDA uses ultrasound machines to look inside cattle and see how much fat is on each animal. Cattle with more fat make steaks that taste better and are more tender.

Thermometers

Thermometers measure temperature, something that affects how quickly food spoils. USDA inspectors check thermometers used in transport trucks to make sure food is being shipped safely. They also monitor thermometers in restaurants and stores to ensure that foods are being kept at safe temperatures.

SMSs

Soil moisture sensors (SMSs) measure how much water is in the soil. USDA crop scientists rely on SMSs because different moisture levels help different crops grow. Some crops need more water than others. SMSs help farmers decide what crops to plant and where.

Microscopes

USDA scientists use microscopes to study plants and animals. These tools magnify objects, such as pieces of tissue. Tissue samples are sometimes treated with chemical stains. This makes certain parts of the tissue easier to see so that scientists can more easily examine certain areas of it.

USDA in the United States

The USDA has offices in every U.S. state. Every state produces food of some kind. Some states grow more fruits and vegetables. Others have large amounts of beef cattle or poultry.

1

Washington, DC

The USDA's headquarters are located in Washington, DC. The main office is in the Jamie L. Whitten Building. It faces the National Mall.

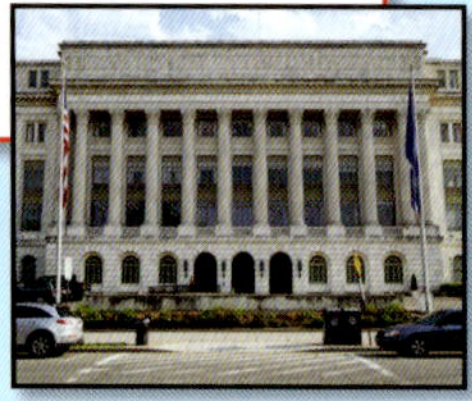

2

Salinas, California

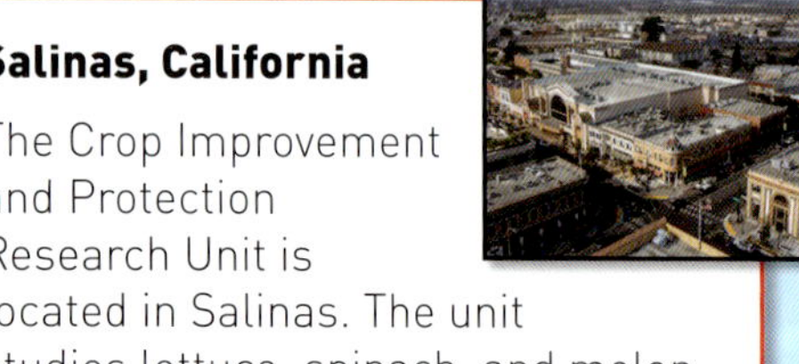

The Crop Improvement and Protection Research Unit is located in Salinas. The unit studies lettuce, spinach, and melon crops. By doing so, it identifies problems that cause diseases. This information is then used to help farmers control weeds and protect the health of their crops.

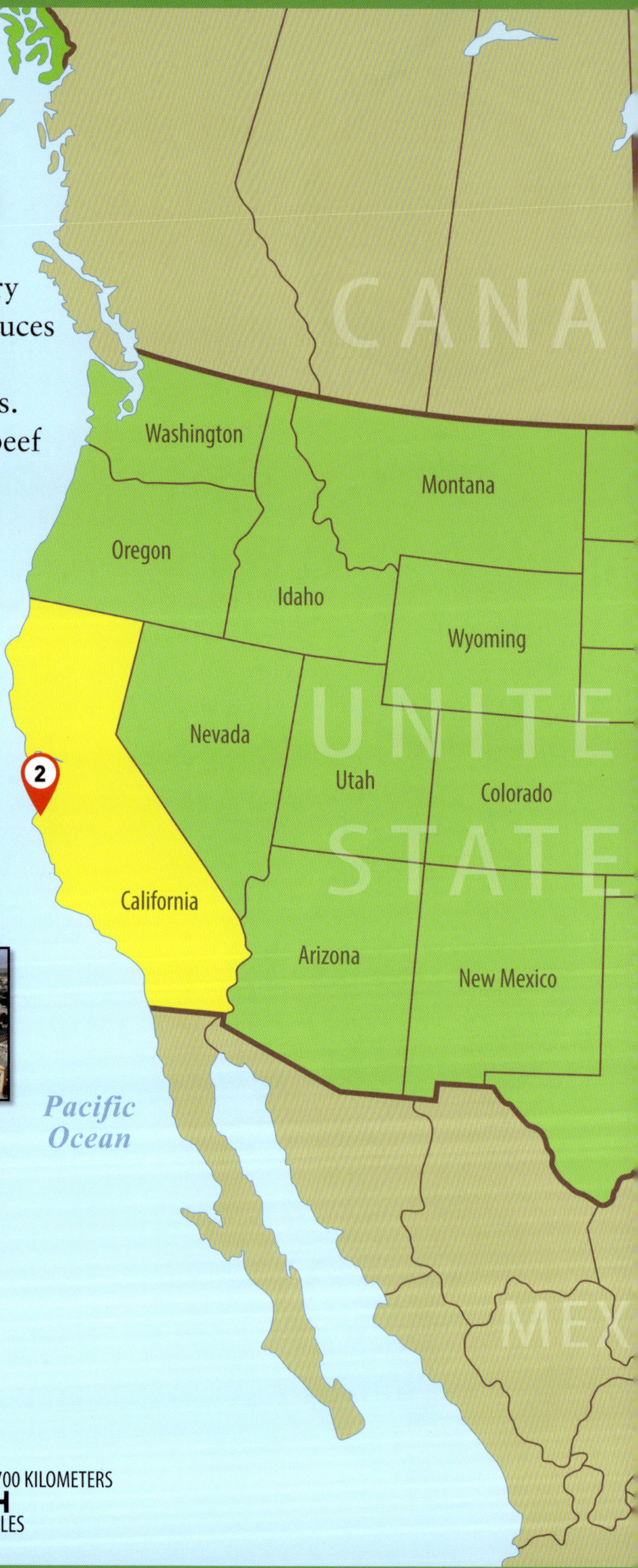

LEGEND
- Land (USA)
- Land (Other)
- Water

SCALE
700 KILOMETERS
350 MILES

3

Beltsville, Maryland

The Beltsville Agricultural Research Center was created in 1910. Scientists at the center study crops and food animals. They research better ways to grow food. Their work helps farmers both in the United States and around the world.

4

Stoneville, Mississippi

Scientists with the Crop Genetics Research Unit, in Stoneville, study cotton and soybean plants. They compare different types of these plants in order to develop crops that are easier to grow. Some varieties resist disease. Others need less water to grow.

USDA in the World

The United States exports a range of unprocessed foods. These include grains, tree nuts, and livestock. The country also exports many processed foods, such as drinks and baked goods. The United States is the world's largest exporter of food and agricultural products. The Foreign Agricultural Service (FAS) is an agency of the USDA. It helps the United States sell this food, along with other agricultural products, to countries around the world.

China is the number-one **importer** of U.S.-grown food, followed by Mexico. Canada is another major U.S. agricultural market. Most of the fruits and vegetables the country imports, such as lettuce and oranges, are grown in the United States. India is another destination for many U.S. foods.

The FAS has nearly 100 international offices around the world. This allows the agency to help advance U.S. agricultural trade and improve food security in many different countries, including Pakistan.

India and China are two of the most-populated countries on Earth. Due to their large populations, these nations rely on imports to feed their people. Thanks to USDA research, the United States grows almost as much food as China on less land and with fewer workers, allowing it to export this food. The USDA exports this research, helping farmers in other countries grow more food more efficiently.

The FAS was founded in 1953. It replaced the Office of Foreign Agricultural Relations (OFAR).

FOREIGN AGRICULTURAL SERVICE

USDA

The **United States** and **Canada** are equal **partners** in the food trade. In **2019**, U.S. food and agricultural product exports to and imports from Canada were each **$24 billion**.

U.S. soybean yields have **doubled** since **1948**. **Corn** yields have **quadrupled** over the same period of time.

The **USDA's goal** is to **increase** U.S. **agriculture production** by **40 percent** between **2020** and **2050**.

USDA Today

President Abraham Lincoln called the USDA "The People's Department." That nickname still applies to this day. The USDA touches the life of every U.S. citizen, every day. While not all people make a living growing food, they all need to eat. Every American requires a safe supply of food.

U.S. farms and ranches create millions of jobs. Many of these jobs are related to food. Some pack it or cook it. Others taste food for a living. The USDA's regulations help ensure that these people are properly trained and do their jobs in a way that keeps both them and the people who eat the food they handle safe.

Agricultural producers grow more than food. They also grow plants used to make textile products, such as clothing. Today, crops that are used to make plant-based fuel, which can be used in place of oil and gas, have also become important. The industry produces leather and ingredients used in medicine as well. The USDA is responsible for supporting these producers while ensuring that they go about their business safely and effectively.

In recent years, the USDA has worked with other federal agencies, including the Federal Emergency Management Agency (FEMA), to try to reduce wildfires in important forests and grasslands across the United States.

Food Safety

MODERN CASE STUDY

In 1993, hundreds of U.S. citizens ate meat infected with *E. coli* bacteria and became sick. Four children, in Washington and California, died. The meat was served at a chain of fast-food restaurants. In the aftermath, people were worried about the safety of the country's food supply. They called for stricter food safety rules.

The USDA studied the problem. Scientists worked with companies that sold meat to restaurants. They also looked at how employees at restaurants handled uncooked meat. In 1996, the USDA changed its rules for meat and egg inspections. Inspectors now specifically looked for bacteria such as *E. coli*. The USDA also required several industries to follow new rules. These rules applied to meat packing plants, food factories, and grocery stores. Cooks working in restaurants were also required to follow these rules.

Today, the USDA tells people when a food product might not be safe to eat. It is able to use tools such as social media to spread information and alerts about possible contaminated foods.

E. coli is often harmless. The bacteria can make people sick if it mutates in food that is then eaten.

Before 1996, food inspectors in the United States mostly tried to find contaminated meat simply by touching, tasting, and smelling it.

USDA Looking to the Future

Moving forward, the USDA will need to balance the needs of many different groups. For example, farmers and food businesses need to be able to make money. Consumers need to know more about how their food is grown. They want healthy and affordable food.

One USDA goal is to increase U.S. food production. Most of that increase will come from growing more plants. Some of these will be used to feed livestock. Others will produce fiber. Most states now allow the production of hemp, a high-value fiber that can be used to make cloth. The USDA is also moving to promote plant-based fuels.

In 2021, the United States exported 1.2 billion gallons (4.5 billion liters) of ethanol, a fuel made from corn.

Today, about one-third of U.S. farmers and ranchers are more than 65 years of age. Farms and ranches are also bigger than ever. That means fewer and older people now own more of the land used to grow food. In coming years, the USDA may need to find new ways to help encourage and support young farmers to keep the agriculture industry of the United States healthy and strong.

Almost 90 percent of farms in the United States are small. However, large family farms, which make up about 3 percent of the country's total farms, are responsible for 46 percent of the nation's production.

The number of farms in the United States has steadily dropped since the 1930s.

ACTIVITY ★★

Create a Policy Paper

The USDA works for the people of the United States. The department protects food and jobs. The food industry depends on a healthy environment. To grow food, it needs clean air and water. It needs healthy soil. These are natural resources. To be sustainable, the industry must safeguard such resources.

Develop your own thoughts on how much attention the USDA should give to environmental issues. Write a policy paper that summarizes your opinions.

Step 1:

Answer the following questions to help you develop your opinions.

1. Should the USDA make the environment a priority? Are farm profits more important than the environment? Why or why not, and, to what extent?
2. Should the USDA look for areas where food and agriculture production improve the environment? Why or why not?
3. Should the USDA draw attention to practices that harm the environment? Why or why not?
4. Who should be responsible for ensuring the right to clean air and water is met? Why?
5. Who should have to take care of the soil food growers use to grow food? Why?
6. Does the United States have the right to make its citizens farm their land in a certain way? Why or why not?
7. Do you think all food and agriculture producers should have to obey the same sets of laws? Why or why not?

Step 2:

Take your opinions from questions 1 to 7 and write a one-page policy paper. This should explain to what extent the USDA should focus on environmental issues. Follow the format outlined below.

- Paragraph 1: What is the problem?
- Paragraph 2: What steps can solve the problem?
- Paragraph 3: What is your policy on the issue, and why?

QUIZ ★★

1 What is the value of U.S. food and agricultural products exported each year?

2 Which four states grow the most food in the country?

3 Which president created the USDA in 1862?

4 In what year did the USDA create its first website?

5 What did President Lincoln call the USDA?

6 Who was the first woman appointed secretary of the USDA?

7 What percentage of the U.S. population lives on farms and ranches?

8 In what year did the USDA open the Bureau of Home Economics?

9 How many people received assistance from SNAP in 2021?

10 What portion of U.S. farmers and ranchers are older than 65?

ANSWERS

1. About $139 billion 2. California, Iowa, Nebraska, and Texas 3. Abraham Lincoln 4. 1995 5. "The People's Department" 6. Ann Veneman 7. Less than 2 percent 8. 1923 9. About 41.5 million 10. One-third

KEY WORDS ★★

agrarian: people who live on farms and make their living from growing food

budget: a plan for spending money

civil rights: the liberties and freedoms guaranteed to citizens of the United States

constitution: a country's basic laws, which state the rights of the people and the powers of the government

executive branch: the U.S. president, vice president, and cabinet

export: sale of a product to another country

food-insecure: unable to easily afford or acquire enough healthy food

genetically modified: a living thing that has been altered to have certain traits or characteristics

Great Depression: a global reduction in economic activity that began in 1929. It included widespread drought and job loss.

importer: a nation that brings in products from another nation

judicial branch: the part of the U.S. government made up of the judges and justices who interpret laws and decide if they are constitutional

legislative branch: the U.S. Congress, composed of the Senate and the House of Representatives

loans: money lent that is expected to be paid back over time

nutrition: consuming and using food

organic: in food production, this refers to food produced without artificial chemicals

rural: sparsely populated areas outside of large communities

INDEX ★★

Published by Lightbox Learning Inc.
276 5th Avenue, Suite 704 #917
New York, NY 10001
Website: www.openlightbox.com

Library of Congress Cataloging-in-Publication Data

Names: Gregory, Joy, author.
Title: Department of Agriculture / Joy Gregory.
Description: New York, NY : Smartbook Media Inc., [2022] | Series: Power, authority, and governance | Includes index. | Audience: Grades 4-6
Identifiers: LCCN 2021033732 (print) | LCCN 2021033733 (ebook) | ISBN 9781510558786 (library binding) | ISBN 9781510558793
Subjects: LCSH: United States. Department of Agriculture--Juvenile literature. | Agriculture and state--United States--Juvenile literature.
Classification: LCC S21.C9 G77 2022 (print) | LCC S21.C9 (ebook) | DDC 354.5/0973--dc23
LC record available at https://lccn.loc.gov/2021033732
LC ebook record available at https://lccn.loc.gov/2021033733

Printed in Guangzhou, China
1 2 3 4 5 6 7 8 9 0 27 26 25 24 23

012023
111121

Art Director: Terry Paulhus Project Coordinator: John Willis

The publisher acknowledges Alamy, Getty Images, Newscom, Shutterstock, and Wikimedia as the primary image suppliers for this title.